Amer Taqa

# Some Complexes of N-Aryl Furfural Nitrones with Co(II), Ni(II), Cu(II), Zn(II) and Cd(II) Chlorides

Der GRIN Verlag publiziert seit 1998 wissenschaftliche Arbeiten von Studenten, Hochschullehrern und anderen Akademikern als eBook und gedrucktes Buch. Die Verlagswebsite www.grin.com ist die ideale Plattform zur Veröffentlichung von Hausarbeiten, Abschlussarbeiten, wissenschaftlichen Aufsätzen, Dissertationen und Fachbüchern.

**Document Nr. V209690**

**Amer Taqa**

# Some Complexes of N-Aryl Furfural Nitrones with Co(II), Ni(II), Cu(II), Zn(II) and Cd(II) Chlorides

GRIN Verlag

Die Deutsche Bibliothek verzeichnet diese Publikation in der Deutschen Nationalbibliografie; detaillierte bibliografische Daten sind im Internet über http://dnb.d-nb.de/ abrufbar.



**1. Auflage 2011**
**Copyright © 2011 GRIN Verlag GmbH**
**http://www.grin.com**
**Druck und Bindung: Books on Demand GmbH, Norderstedt Germany**
**ISBN 978-3-656-40083-7**

# SOME COMPLEXES OF N-ARYL FURFURAL NITRONES WITH Co(II),Ni(II), Cu(II),Zn(II) and Cd(II)  CHLORIDES


## Amer A. Taqa

Department of Dental Basic Science/ College of Dentistry/ University of Mosul



Abstract

Some new metal(II) dichloride complexes with the ligands substituted nitrones of the general formula [ML$_2$Cl$_2$], where M= Co(II), Ni(II), Cu(II), Zn(II) and Cd(II), L=OCH=CHCH=C-CH=N(O)C$_6$H$_4$X  (X=H,p-CH$_3$,CH$_3$O,CH$_3$CO,F,Cl,and  Br)  have been prepared and characterized  by elemental analysis, IR,$^1$H,$^{13}$C  NMR and Vis/Uv spectroscopy. The IR spectral data showed that the  nitrone ligands coordinated with the metal ion through the most active atom of the N-oxide to give square planner coordinate (Cu,Ni,) complexes and (Zn,Cd,Co)  tetrahedral complexes. No correlation was observed between the N-O vibrations stretching hing frequency ν(N-O) of the complexes and the Hammet (σ) constants.


## Introduction

Preparation of nitrone compounds and it`s derivatives had much attention because of the biological importance, new methods of prepared nitrone compounds have much attention [1-2].  Some of these methods have been applied to the preparation of complexes molecules with useful biological activity such as antibiotics and glycosides inhibitors [3-8]. Therefore, new methods of activation, such as microwave chemistry and coordination to a metal center, have been attempt.  In fact, it was observed that the microwave field decreased the activation energy of various types of reactions in particular with organonitrones [9].Moreover, metals in coordination processes can dramatically increased the reactivity of organonitriles [10, 11] and metal-mediated processes can lead to the formation of heterocyclic species [10].

Nitrones have been recognized as having the following resonance structure

When A,A` or B is a mesmeric substituent the nitrone group (-C=NO-) will, of course, interact mesomerrically with the substituent and thereby will probably exert an electron-attracting effect on the latter.

The $^{13}$CNMR and $^{1}$H1NMR study showed [12], when a strong electron-donating group is in the Para position of a phenyl ring the nitrone acted as an electron-withdrawing group. For strong electron-withdrawing substituent the nitrone group acted as an electron donor. Thus they concluded that the nitrone group behaved as an electron-withdrawing or donating

In the current work, we have presented the reaction between metal chloride Co(II), Ni(II), Cu(II),Zn(II),Cd(II) and N-arylfurfuralnitrones(fig1) in order to examine the type of interaction between metals and this type of ligands as they contain more than possible donor site. Also measure the effect of substituent on the mechanism formation of complexes using Hammet equation. To the best of our knowledge, this work is novel.

## EXPERIMENTAL

The ligands p-x-phenyl-N-furfural nitrones (X=H, Cl, Br, F, OH, $CH_3$, $OCH_3$ and $COCH_3$) were prepared as described previously with substituted $X-C_4H_4NHOH$ in ethanol [13]. The free nitrone ligands were purified by crystallization.

Preparation of complexes: The following standard method was used; molar quantities (usually 1:2 mmole) of metal salts and the nitrone ligand (L) were dissolved in absolute ethanol (25ml) at ambient temperature. The colored solution and the complex started to deposit. The formed precipitate were filtered, washed several times with small portions of ether and dried. The yield is almost quantitative.

Metal analysis for some of the complexes were determined, Nickel metal was determined as dimethylglyoxime complexes [14]. Cobalt, zinc, copper and cadmium metals were determined by pyridine method [14].

$^{13}$C and $^{1}$H1NMR spectra were recorded at 25$^{\circ}$C on a Bruker DPX 300MHz spectrometer at the department of Chemistry, College of Science, Jordan.

IR spectra was recorded on an Infrared spectrophotometer BRUKER (TENSOR 27), conductivity measurements were done for $10^{-3}$M solutions of the complexes in ethanol and dimethylforamide at room temperature (25$^{\circ}$C), using a Jenway conductivities meter model 4070. Visible spectra were recorded Ultraviolet–Visible spectrophotometer (UV -1650 PC).

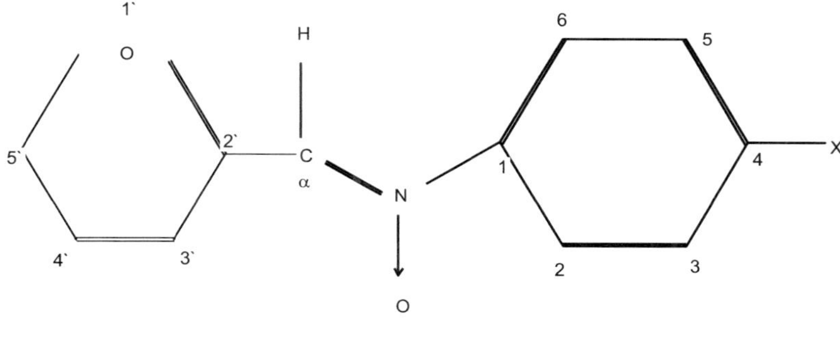



L₁: X=H

$L_1$: X=H

$L_2$: X=CH$_3$

$L_3$: X=OCH$_3$

$L_4$: X=COCH$_3$

$L_5$: X=F

$L_6$: X=Cl

$L_7$: X=Br

Fig.1. The Nitrones Ligands used in the coordination with $MCl_2$(M= Mn,Co,Ni,Cu,Zn and Cd )

## Results and Discussion

The IR data in Table I confirmed the formation of the complexes. There are changes in the frequencies of the C=N band upon complexation, but especially significant is the appearance of a new band at ca. 340 cm$^{-1}$, attributed to $\upsilon$(M-O), which served as a good indicator of coordination.[15]. Moreover, the drastic shift in the $\upsilon$(N-O) frequency is clear evidence for the interaction between the NO group of the nitrone ligand and metal. However, coordinatid lead to a shift to lower frequency and the values of v(NO)$_{complex}$- v (NO)$_{ligand}$ showed a systematic variations from 15-to-50 cm$^{-1}$ (Table1), and this may be attributed to a decrease in the NO group upon coordination[16]. On the other hand the v (C=N) frequency of the ligand show a great change to a higher values upon coordination and the values of v (CN)$_{complex}$- v (CN)$_{ligand}$ show again systematic variations from 25 to 45cm$^{-1}$. This may be due to the increase in the bond order between C and N upon coordination. In

contrast, the ν (C-O) frequency of the ligand which appeared in the range [17] 1070-1090cm$^{-1}$ remains almost constant upon coordination supported that the furfural oxygen remained unchenged upon coordination supported that the furfural oxygen is not involved in the coordination.

*Molar conductivities*

Molar conductivities for $10^{-3}$M solutions of the complexes in two different solvents, ethanol and DMF, at ambient temperature 25°C were in the range 1.3-11.82 and 0.05-10.2 ohm cm$^2$ mole$^{-1}$ respectively,(Table II) suggesting the present of non-conductive species[18](*i.e.,* non-ionic) complexes in the solvents used.

NMR Spectra

The magnetic resonance (Table III) showed shifted α-H of ligand to complex in $d_6$DMSO solvent and this confirms coordination. The $^{13}$C NMR spectral data were recorded to provide an additional indicator for the coordination number. In (Table IV), shows a clear change in the chemical shifts of the carbon atoms of the organic nucleus, especially significant are those of C-α, C-1, C2□`,C3`and C5`(Fig1). Carbon atoms were assigned by comparison with other related organic compounds as model compounds[19]. The carbon C-α showed a downfield shift on going from the free ligand to its complex (*ca.*1.5ppm) and this clearly suggested that the C=N bond order is increased due to complexiation. This is supported by the stretching frequency of the C=N bond which shifted to a higher value on complexation. Similarly, C-1 is also affected upon coordination and is shifted upfield by *ca.* 1ppm. Although the furfural nucleus had not been involved in the coordination, nevertheless, the peak for C-2` is shifted upfield whereas those of C-3` and C-5` are shifted downfield upon coordination. We believe that this may be due to the inductive effect caused by coordination. This in turn causes a great drainage of electron density from the furfural oxygen, through conjugation to C2`, then to C-α and so on.

From Fig1. It can be seen that only two active donor site, i.e., the furfural oxygen and the nitrone oxygen, participate in bonding. Whereas the ligands L3 and L4 contain further donor site, i.e., the methoxy oxygen and the acetoxy oxygen atoms, respectively. Nevertheless, the spectral data showed that both groups, the methoxy and the acetoxy, were not involved in the complexation with metal. Therefore, the only possible donor sites of all ligands are the furfural oxygen and the nitrone oxygen atoms, thus the nitrones can behave as bidentate ligands, since the furfural group can rotate freely around the C2-Cα bond, and both oxygens can be arranged in such a way that they can complex with metal in a bidentate fashion. However, this is not the case with complexes, in which the IR spectral data showd that all of these ligands coordinate with metal in a bidentate fashion *via* both oxygen atoms[16,20].

The electronic spectra of the copper(II) complexes showed a single broad and poorly defined asymmetric band around 17500 cm$^{-1}$ and the spectra of nickle(II) Complexes showed a band around 20000cm$^{-1}$. These results were consistent with square-planar structures, since the four lower orbitals are often so close together in energy, that individual transitions there from to the upper d level cannot be distinguished – hence the single absorption band[21].

The nickel (II) complexes showed a diamagnetic behavior consistent with square-planar environment around the metal ion. Magnetic susceptibilities of the copper (II)

complexes were 1.87 BM. The effective magnetic moments were in accordance with diluted monomeric units [22].

From these data, square-planar structures may be proposed for the nickel(II) and copper(II) complexes with the monodentate ligand and the fourth coordination position occupied by two chloride ions, as depicted in fig.2; A similar structure but tetrahedral may be suggested for the zinc(II) and cadmium(II) complexes.

Studies of the magnetic and spectral properties of the prepared Co (II) complexes gave $\mu_{eff}$ values (3.59-3.7 BM) and one electronic spectral band in the visible spectrum at (14850, 16100 cm$^{-1}$) which assignment to $^4A_2 \longrightarrow {}^4T_{1(p)}$ . The magnetic moment of cobalt (II) indicates the presence of three unpaired electrons. These values suggest the geometry of tetrahedral configuration. The suggestion were confirmed by a band in electronic spectra which is assigned to ($^4A_2 \longrightarrow {}^4T_{1(P)}$) transition [24].

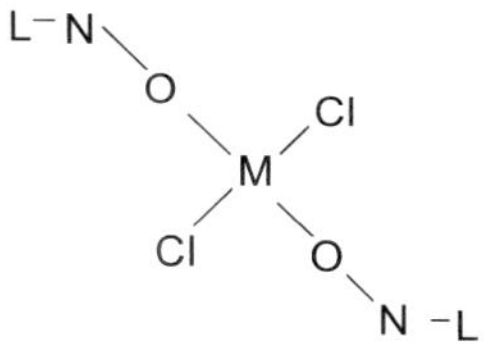

L⁻N
O
Cl
M
Cl
O
N ⁻L

Fig. 2. The Suggested Structure for the MCl$_2$.L$_2$ Complexes

(for M and L see Fig. 1)

No correlation were observed between the N-O vibration stretching frequency $\nu$ (N-O) of the complexes with the Hammet ($\sigma$) constants (Fig.3) i.e. the plot of $\nu$ (N-O) for a given measurements with sigma constant of Para substituent constant [25] will not give a straight line, it is likely that the mechanism of the formation of complexes changed upon adding a different substituent. Other deviations from linearity may be due to a change in the position of the transition state. In such a situation, certain substituent may be caused the transition state to appear earlier (or later) in the reaction mechanism [26].

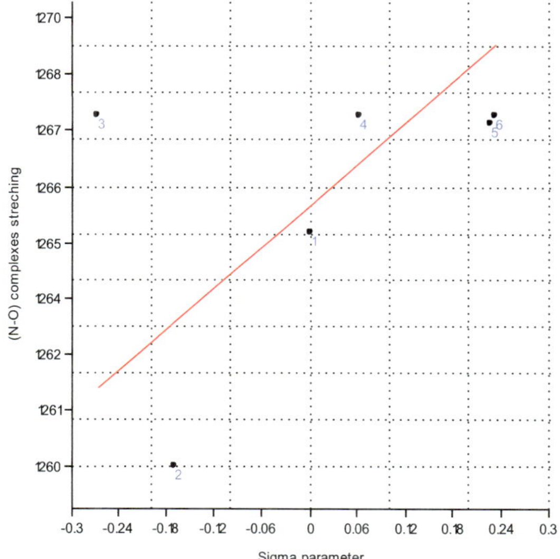


Figure 3. The correlation between the νN-O of complexes and Hammet constant σ .

## References


1.Januario M.A..C.,Vadim Y.K. and Armando J.L .Pomberio, Dalton trans. ,2540-2543 (2003).

2  Murielle C., Michael R.,Francoise R.M.,Eric R. and Sandrine P., Chem..Comun. 2330-2331(2004).

3.Z.Guo and P.J.Sadler, Adv.Inorg.Chem., 49,183,(1999).

4. Z.Guo and P.J.Sadler, Angew. Chem., Int. Ed., 38, 1512,(1999).

5.K.V. Gothelf and K. A.Jorgenson,Chem.Rev., 98,863,( 1998).

6.T.A.K.Allaf, R.I.H. Al-Bayati, L.J.Rashan and R.F.Khuzaie, Appl. Organomet., Chem.,10,47,(1996).

7. M.A. J.Charmier, V.Y.Kukushkin and A.J.L.Pombiro, Dalton Trans.,2540-2543,(2003).

8.A.K.Saxena and F . Huber, Coord. Chem. Rev.,95,109,(1989).

9. S.Caddick, Tetrahedron, 51,10403;(1995).

10.V .Yu, Kukush kin and A.J.L. Pombeiro, Chem. Rev., 102,1771,(2002).

11.. R.A.Michelin ,M.Mozzen and R.Bertani, Coord.Chem. Rev., 147,299,(1996).

12.Narumugam,  P.Manisanker and S.Sivasubramanian and D.A.Wilson., Organic
   Magnetic  Resonance, Vol,22,9, 592-596,(1984).

13. N.Arumuga m, P.Manisanker,S.Sivasubramanian and D.A.Wilson, Org. Mag.
   Res.,22,592,( 1984).

14. A.I.Vogel "Text book of practical organic chemistry" Third edition Logman group
   Ltd.  P.863, (1972).

15.T.A.K.Al- Allaf and  M.A.Al-Tayy, J. Organometal. Chem., 391,37(1990); and
   references  therein.

16. T.A.K.A l-Allaf  and A.O.Omer, Dirasat(Jordan), 20B,53(1993) ;Chem. Abst. 120
    217910( 1994).

17. R.M.Sil verstein, C.G. Basker and T.C. Morrill, "Spectrometric Identification of
   Organic  Compounds", J.Wiley and Sons, New York, 3$^{rd}$ Ed.(1974).

18. S.F.A. Kettle, "Coordination Compounds" Thomas Nelson, London(1975).

19. G.C.Levy, R.L. Lichter and G.L.Nelson, "Carbon-13 Nuclear Magnetic
   Resonance Spectroscopy" J.Wiley and Sons, New York (1980).

20. T.A.K .Al-Allaf, J.M. Al-Rawi and A.O.Omer, Iraqi  J. Chem., 15,22(1990).

21. Lever ,A.B.P. Inorganic Electronic Spectroscopy, 2$^{nd}$ ed., Elsevier, Amsterdam,
   (1984 ).

22. Carl in, R.L. Magnetochemistry, Springer-Verlag, Berlin, (1986).

23. Nicholls. " Complexes and first –row transition elements" White friars press
   LT D, (1974), P.1,100

24. W .D.Honnick and J.J.Zuckerman, "Organotion (II)-oxygen and – sulfur H
   etrocycles through protolysis of Tin(II) dimethoxide"  Inorg.Chem.,17, pp. 501
   5 04,(1978).

25 . C.Hansh, A.Leo and R.W. Taft Chem.Rev.;91;165-195,(1991).

2 6.E.V.Anslyn,D.A.Dougherty, Modern Physical Organic Chemistry. University
   Science Books ISBN 1-891389-31-9.

Table I:  Physical properties and infrared spectra for complexes.

| Complex | Color | Empirical Formula | Formula Weight | Yield (%) | mp°C | $\nu$(N-O) | $\nu$ (C=N) | $\nu$ (M-O) | $\nu$ (M-Cl) |
|---|---|---|---|---|---|---|---|---|---|
| $CoCl_2.L_1$ | D.red | $C_{22}H_{18}O_4N_2CoCl_2$ | 504.23 | 90 | >350 | 1260m | 1625m | 335w | 250s |
| $NiCl_2.L_1$ | red | $C_{22}H_{18}O_4N_2NiCl_2$ | 503.995 | 90 | >350 | 1270s | 1620m | 350m | 260m |
| $CuCl_2.L_1$ | green | $C_{22}H_{18}O_4N_2CuCl_2$ | 508.847 | 90 | d306 | 1260m | 1620m | 350m | 260s |
| $ZnCl_2.L_1$ | white | $C_{22}H_{18}O_4N_2ZnCl_2$ | 510.691 | 90 | >350 | 1265m | 1610m | 345w | 260m |
| $CdCl_2.L_1$ | whit | $C_{22}H_{18}O_4N_2CdCl_2$ | 557.712 | 90 | >350 | 1270s | 1610m | 340w | 265m |
| $CoCl_2.L_2$ | red | $C_{24}H_{22}O_4N_2CoCl_2$ | 532.288 | 85 | 290d | 1270s | 1610m | 350m | 270s |
| $NiCl_2.L_2$ | Brown | $C_{24}H_{22}O_4N_2NiCl_2$ | 532.049 | 95 | >350 | 1270s | 1620m | 350m | 260m |
| $CuCl_2.L_2$ | green | $C_{24}H_{22}O_4N_2CuCl_2$ | 536.901 | 90 | 290d | 1265m | 1620m | 340w | 235w |
| $ZnCl_2.L_2$ | light white | $C_{24}H_{22}O_4N_2ZnCl_2$ | 538.745 | 90 | 310d | 1260m | 1625m | 340w | 255w |
| $CdCl_2.L_2$ | white | $C_{24}H_{22}O_4N_2CdCl_2$ | 585.766 | 94 | 266d | 1270s | 1620m | 335w | 245w |
| $CoCl_2.L_3$ | red | $C_{24}H_{22}O_6N_2CoCl_2$ | 564.287 | 40 | >350 | 1275m | 1625m | 350w | 250s |
| $NiCl_2.L_3$ | red | $C_{24}H_{22}O_6N_2NiCl_2$ | 564.047 | 90 | >350 | 1260s | 1628m | 350m | 250m |
| $CuCl_2.L_3$ | green | $C_{24}H_{22}O_6N_2CuCl_2$ | 568.9 | 70 | 189d | 1270m | 1615m | 335w | 270s |
| $ZnCl_2.L_3$ | white | $C_{24}H_{22}O_6N_2ZnCl_2$ | 570.744 | 65 | >350 | 1270s | 1625m | 340w | 243w |
| $CdCl_2.L_3$ | white | $C_{24}H_{22}O_6N_2CdCl_2$ | 617.765 | 86 | >350 | 1265m | 1630m | 340w | 235w |
| $CoCl_2.L_4$ | red | $C_{26}H_{22}O_6N_2CoCl_2$ | 588.309 | 85 | 247 | 1270s | 1615m | 340w | 255s |
| $NiCl_2.L_4$ | red | $C_{26}H_{22}O_6N_2NiCl_2$ | 588.069 | 90 | >350 | 1270s | 1610m | 355w | 250m |
| $CuCl_2.L_4$ | Dark green | $C_{26}H_{22}O_6N_2CuCl_2$ | 592.922 | 88 | 300d | 1280b | 1610m | 355w | 265s |
| $ZnCl_2.L_4$ | white | $C_{26}H_{22}O_6N_2ZnCl_2$ | 594.766 | 69 | 278-290d | 1270m | 1620m | 335m | 250s |
| $CdCl_2.L_4$ | white | $C_{26}H_{22}O_6N_2CdCl_2$ | 641.787 | 89 | 350d | 1275s | 1625m | 350w | 250s |
| $CoCl_2.L_5$ | Pale red | $C_{22}H_{16}O_4N_2CoCl_2F_2$ | 540.216 | 96 | >350 | 1270s | 1620m | 340w | 270m |
| $NiCl_2.L_5$ | Red | $C_{22}H_{16}O_4N_2NiCl_2 F_2$ | 539.976 | 90 | >350 | 1265s | 1620m | 340w | 270m |
| $CuCl_2.L_5$ | Green | $C_{22}H_{16}O_4N_2CuCl_2 F_2$ | 544.828 | 90 | 310d | 1270s | 1615m | 340w | 270m |
| $ZnCl_2.L_5$ | White | $C_{22}H_{16}O_4N_2ZnCl_2 F_2$ | 546.672 | 78 | 289d | 1270m | 1620m | 350w | 270m |
| $CdCl_2.L_5$ | White | $C_{22}H_{16}O_4N_2CdCl_2 F_2$ | 593.693 | 68 | 218d | 1265m | 1625m | 350w | 255s |
| $CoCl_2.L_6$ | Red | $C_{22}H_{16}O_4N_2CoCl_4$ | 573.124 | 80 | 290 | 1265m | 1610m | 355w | 255w |
| $NiCl_2.L_6$ | Red | $C_{22}H_{16}O_4N_2NiCl_4$ | 572.884 | 90 | 317 | 1265m | 1615m | 355w | 255w |
| $CuCl_2.L_6$ | Green | $C_{22}H_{16}O_4N_2CuCl_4$ | 577.737 | 79 | 289 | 1265m | 1620m | 355m | 255w |
| $ZnCl_2.L_6$ | White | $C_{22}H_{16}O_4N_2ZnCl_4$ | 579.581 | 95 | >350 | 1260m | 1620m | 340m | 240s |
| $CdCl_2.L_6$ | white | $C_{22}H_{16}O_4N_2CdCl_4$ | 626.602 | 89 | 353d | 1260m | 1620m | 340m | 270w |
| $CoCl_2.L_7$ | Red | $C_{22}H_{16}O_4N_2CoCl_2Br_2$ | 662.027 | 68 | 216 | 1265m | 1620m | 355w | 270w |
| $NiCl_2.L_7$ | Red | $C_{22}H_{16}O_4N_2NiCl_2 Br_2$ | 661.787 | 70 | 327d | 1265m | 1620m | 355w | 255s |
| $CuCl_2.L_7$ | Pale green | $C_{22}H_{16}O_4N_2CuCl_2 Br_2$ | 666.64 | 87 | >350 | 1265m | 1610m | 355w | 270w |
| $ZnCl_2.L_7$ | White | $C_{22}H_{16}O_4N_2ZnCl_2 Br_2$ | 668.484 | 78 | 268 | 1270s | 1610m | 355w | 270w |
| $CdCl_2.L_7$ | White | $C_{22}H_{16}O_4N_2CdCl_2 Br_2$ | 715.505 | 96 | 219 | 1270s | 1615m | 355w | 270w |

d :  decompose, m:medium, s:strong, w:weak

Table II: HCN, metal  analysis and conductivity value for the complexes.

| Complex | H | C | N | Metal % | Conductivity in DMF | Conductivity in ethanol |
|---|---|---|---|---|---|---|
| $CoCl_2.L_1$ | 3.598(3.55) | 52.404 (52.50) | 5.555(5.71) | 11.687(11.91) | 4.92 | 5.33 |
| $NiCl_2.L_1$ | 3.599(3.61) | 52.429(52.399) | 5.558(5.62) | 11.645(11.8) | 8.30 | 6.16 |
| $CuCl_2.L_1$ | 3.565(3.67) | 51.929(52.21) | 5.505(6.01) | 12.48(12.22) | 4.44 | 1.56 |
| $ZnCl_2.L_1$ | 3.552(3.59) | 51.742(51.98) | 5.485(5.66) | 12.08(12.35) | 5.67 | 5.99 |
| $CdCl_2.L_1$ | 3.253(3.19) | 47.379(47.16) | 5.022(5.15) | 20.155(20.21) | 2.38 | 8.21 |
| $CoCl_2.L_2$ | 4.165(3.99) | 54.155(54.21) | 5.262(5.41) | 11.071(11.21) | 0.05 | 6.34 |
| $NiCl_2.L_2$ | 4.167(4.16) | 54.180(54.20) | 5.265(5.26) | 11.031(10.99) | 10.2 | 6.16 |
| $CuCl_2.L_2$ | 4.130(4.16) | 53.690(53.72) | 5.217(5.21) | 11.835(11.770 | 10.02 | 4.70 |
| $ZnCl_2.L_2$ | 4.115(4.11) | 53.506(53.62) | 5.199(5. 23) | 12.137(11.98) | 4.23 | 6.44 |
| $CdCl_2.L_2$ | 3.785(3.77) | 49.211(50.01) | 4.782(7.77) | 19.190(20.32) | 5.30 | 1.30 |
| $CoCl_2.L_3$ | 3.745(3.77) | 48.687(48.75) | 4.731(4.57) | 10.443(11.00) | 5.5 | 4.99 |
| $NiCl_2.L_3$ | 3.931(3.66) | 51.106(51.00) | 4.966(5.02) | 10.4057(10.41) | 6.01 | 9.00 |
| $CuCl_2.L_3$ | 3.897(4.01) | 50.670(51.01) | 4.924(4.89) | 11.1699(11.32) | 0.09 | 6.99 |
| $ZnCl_2.L_3$ | 3.38(3.16) | 50.50(50.12) | 4.90(4.99) | 11.45(11.55) | 6.12 | 3.21 |
| $CdCl_2.L_3$ | 3.589(3.47) | 46.662(46.81) | 4.534(4.64) | 18.196(18.65) | 5.09 | 3.67 |
| $CoCl_2.L_4$ | 3.769(3.80) | 53.081(53.22) | 4.761(4.75) | 10.017(9.93) | 8.77 | 2.78 |
| $NiCl_2.L_4$ | 3.770(3.79) | 53.103(53.15) | 4.763(4.81) | 9.9806(10.24) | 8.77 | 3.56 |
| $CuCl_2.L_4$ | 3.739(3.69) | 52.668(52.69) | 4.724(4.75) | 10.717(11.01) | 9.04 | 7.66 |
| $ZnCl_2.L_4$ | 3.728(3.76) | 52.505(52.49) | 4.710(4.71) | 10.994(10.68) | 9.91 | 6.90 |
| $CdCl_2.L_4$ | 3.455(3.47) | 48.658(48.69) | 4.364(4.33) | 17.5153(17.58) | 2.56 | 2.78 |
| $CoCl_2.L_5$ | 2.985(3.01) | 48.914(49.11) | 5.1856(5.21) | 10.909(11.23) | 10.0 | 4.98 |
| $NiCl_2.L_5$ | 2.986(3.12) | 48.935(49.04) | 5.187(5.22) | 10.869(11.05) | 8.99 | 9.65 |
| $CuCl_2.L_5$ | 2.960(2.78) | 48.500(48.48) | 5.141(5.19) | 11.6634(11.31) | 6.78 | 7.98 |
| $ZnCl_2.L_5$ | 2.950(3.10) | 48.336(48.54) | 5.124(5.12) | 11.96(12.11) | 9.45 | 4.93 |
| $CdCl_2.L_5$ | 2.716(2.69) | 44.508(45.00) | 4.718(4.73) | 18.934(19.09) | 7.99 | 5.91 |
| $CoCl_2.L_6$ | 2.813(2.82) | 46.105(46.23) | 4.887(4.87) | 10.287(10.51) | 3.33 | 2.56 |

| | | | | | | |
|---|---|---|---|---|---|---|
| NiCl$_2$.L$_6$ | 2.815(2.79) | 46.124(46.21) | 4.889(4.90) | 10.245(10.00) | 3.56 | 3.89 |
| CuCl$_2$.L$_6$ | 2.791(2.79) | 45.737(45.77) | 4.848(4.84) | 10.99(11.43) | 6.98 | 3.878 |
| ZnCl$_2$.L$_6$ | 2.782(2.67) | 45.591(45.62) | 4.833(4.84) | 11.282(11.44) | 3.99 | 5,98 |
| CdCl$_2$.L$_6$ | 3.421(3.45) | 48.185(48.24) | 4.322(4.33) | 17.939(18.31) | 8.99 | 9.85 |
| CoCl$_2$.L$_7$ | 2.43(2.60) | 39.91(42.31) | 4.23(4.48) | 8.91(8.44) | 5.78 | 6.66 |
| NiCl$_2$.L$_7$ | 2.43(2.55) | 39.92(41.77) | 4.23(4.39) | 8.86(8.89) | 6.88 | 7.49 |
| CuCl$_2$.L$_7$ | 2.41(2.83) | 39.63(39.11) | 4.20(4.78) | 9.53(9.21) | 3.99 | 4.98 |
| ZnCl$_2$.L$_7$ | 2.41(2.82) | 39.52(39.23) | 4.19(4.91) | 9.78(9.19) | 8.99 | 8.90 |
| CdCl$_2$.L$_7$ | 2.25(2.79) | 36.93(36.69) | 3.91(4.17) | 15.71(15.21) | 9.34 | 10.9 |

( )found

| Compound | $\delta$ X | $\delta$ H-C=N | $\delta$ H2,6 | $\delta$ H3,5 | $\delta$ H 3` | $\delta$ H4` | $\delta$ H5` |
|---|---|---|---|---|---|---|---|
| $L^1$ | 7.44m | 8.12s | 7.76m | 7.44m | 7.54d | 6.60dd | 7.98d |
| $NiCl_2.L_1$ | 7.55m | 8.12s | 7.72m | 7.50m | J=1.5 | J=1.6 | J=3.5 |
|  |  |  |  |  | 7.65d | 6.70dd | 8.00d |
|  |  |  |  |  | J=1.4 | J=1.6 | J=3.6 |
| $L^2$ | 2.4s | 8.12s | 7.68d | 7.25 d | 7.56 d | 6.63 dd | 7.99 d |
|  |  |  | J=8.4 | J=8.4 | J=1.2 | J=1.5 | J=3.5 |
| $CuCl_2.L_2$ | 2.39 s | 8.05 s | 7.57d d | 7.25 d | 7.60 d | 6.65 dd | 7.96 d |
|  |  |  | J=8.4 | J=8.1 | J=1.3 | J=1.5 | J=3.6 |
| $L^3$ | 3.81 s | 8.05 s | 7.70 d | 6.91 d | 7.53 b | 6.60 dd | 7.93 d |
|  |  |  | J=9.0 | J=9.0 |  | J=1.5 | J=3.4 |
| $CdCl_2.L_3$ | 3.88 s | 8.06 s | 7.71 d | 6.97 d | 7.63b | 6.67 dd | 7.95 d |
|  |  |  | J=8.5 | J=9.0 |  | J=1.6 | J=3.6 |
| $L^4$ | 2.60 s | 8.25 s | 8.09 d | 7.92 d | 7.62 d | 6.67 dd | 8.08 d |
| $ZnCl_2.L_4$ |  |  | J=8.5 | J=8.7 | J=1.3 | J=1.6 | J=3.0 |
|  | 2.65 s | 8.23 s | 8.08 d | 7.90 d | 7.64 d | 6.68 dd | 8.04 d |
|  |  |  | J=8.9 | J=8.6 | J=1.3 | J=1.6 | J=3.6 |
|  | 8.12 s |  |  |  |  |  |  |
| $L^6$ | ------ | 8.14 s | 7.75 d | 7.42 d | 7.57 d | 6.63 dd | 8.00 d |
| $CoCl_2.L_6$ |  |  | J=8.8 | J=8.8 | J=1.5 | J=1.7 | J=3.5 |
|  |  | 8.12s | 7.69 d | 7.46 d | 7.66 d | 6.69 dd | 7.97 d |
|  |  |  | J=8.5 | J=8.5 | J=1.5 | J=1.6 | J=30 |

a Downfield from internal TMS at room temperature using CDCl$_3$ as a solvent. S, single;d,double;dd,doublet of doublets;m,multiple.

b Poorly resolved doublet (broad).

Table IV: Carbon-13 NMR Data[a], $\delta$(ppm) and $J(H_2)$, for Selected Free Ligands and their Metal complexes.

| Compound | $\delta x^c$ | $\delta C\text{-}\alpha$ | $\delta$ C-1 | $\delta C$-2,6 | $\delta C$-3,5 | $\delta$ C-4 | $\delta C$-2` | $\delta C$-3` | $\delta C$-4` | $\delta C$-5` |
|---|---|---|---|---|---|---|---|---|---|---|
| $L^1$ | ———— | 144.7 | 147.6 | 121.1 | 129.2 | 130.0 | 147.3 | 116.5 | 112.7 | 124.3 |
| $CoCl_2.L_1$ | ------- | 146.6 | 146.6 | 121.8 | 129.5 | 130.6 | 146.6 | 118.9 | 113.2 | 128.2 |
| $L^2$ | 21.1 | 144.5 | 147.6 | 120.8 | 129.7 | 140.2 | 145.1 | 116.2 | 112.7 | 123.8 |
| | 21.2 | 146.0 | 146.6 | 121.6 | 129.9 | 141.0 | 144.2 | 118.8 | 113.2 | 127.9 |
| $CuCl_2.L_2$ | | | | | | | | | | |
| $L^3$ | 55.6 | 144.4 | 140.6 | 114.1 | 123.5 | 160.7 | 147.7 | 116.1 | 112.7 | 123.8 |
| | 55.7 | 145.8 | 139.8 | 114.4 | 123.0 | 161.1 | 146.7 | 118.4 | 113.1 | 127.9 |
| $CuCl_2.L_3$ | | | | | | | | | | |
| $L^4$ | 26.8 | 145.3 | 150.0 | 121.2 | 129.4 | 137.9 | 147.4 | 117.7 | 113.0 | 124.9 |
| $ZnCl_2.L_4$ | 26.8 | 146.0 | 149.7 | 121.5 | 129.0 | 138.0 | 147.0 | 118.6 | 113.0 | 126.5 |
| $L^6$ | ———— | 144.9 | 147.4 | 122.3 | 129.3 | 135.7 | 145.6 | 116.9 | 112.8 | 124.2 |
| $NiCl_2.L_6$ | ------- | 146.4 | 146.5 | 123.0 | 129.6 | 136.4 | 144.9 | 119.2 | 113.3 | 128.0 |

a Downfield from internal TMS at room temperature using $CDCl_3$ as a solvent. S, single; d,double; dd,doublet of doublets;m,multiple.

b Poorly resolved doublet (broad).